CHP TECHNOLOGY SERIES

UNDERSEA

Written by
Dan Mackie

Editor:
Janet Stewart

Art:
Rick Rowden, Gary Wein, Tim Zeltner

Second Printing, 1992

DURKIN HAYES PUBLISHING LTD.
3312 Mainway, Burlington, Ontario L7M 1A7, Canada
One Colomba Drive, Niagara Falls, New York 14305, U.S.A.

CONTENTS

ABOUT THIS BOOK

We like to think we know our earth quite well and that the most exciting frontier of the future will be outer space. But the seas around us remain as unexplored as the distant stars. We have only begun to explore the depths of the oceans, partly because it is as difficult to explore the seas as it is to visit outer space. We have to overcome problems of cold, pressure, depth and darkness.

Never before has man had a better opportunity than now to discover what lies beneath the oceans, nor has man had a greater need to do so. We depend on the seas for food, minerals and oxygen. As President Kennedy observed in 1961, "Knowledge of the oceans is more than a matter of curiosity. Our very survival may hinge upon it."

In this book we will take you on a tour of the magical world of our seas and oceans. You will learn how man discovers some of the secrets held by these great bodies of water. You will also come to realize that future explorations are only limited by our imaginations. Excitement and challenge are yours as you begin to explore UNDERSEA.

OCEANS, SEAS, LAKES AND RIVERS

TAKING ANOTHER LOOK AT OUR PLANET

Look at a globe from the center of the Atlantic Ocean: North and South America are on the left, and the land masses of Europe, Africa and Asia stretch into the distance on the right, with Australia below them, surrounded by water. Turn the globe to 165° longitude and we see a different picture. From this position, most of the earth's surface appears covered by water. Now turn the globe and look from the South Pole - water is all around us. In fact, 75% of the earth's surface is covered with water.

Of the approximately 1,000,000 species that inhabit the earth, more than half live in the seas and oceans. While land supports life from the depths of only a few meters to the upper limits of the treetops, the living space of the oceans is much greater. From the surface to the ocean floor thousands of meters below, water supports life in our seas.

We know our land masses very well - we've mapped them all. But mapping the oceans is more difficult. It is only recently, with the aid of modern technology, that man has begun to explore the oceans. And it is only in your lifetime that we have begun to uncover secrets that have been hidden for millions of years.

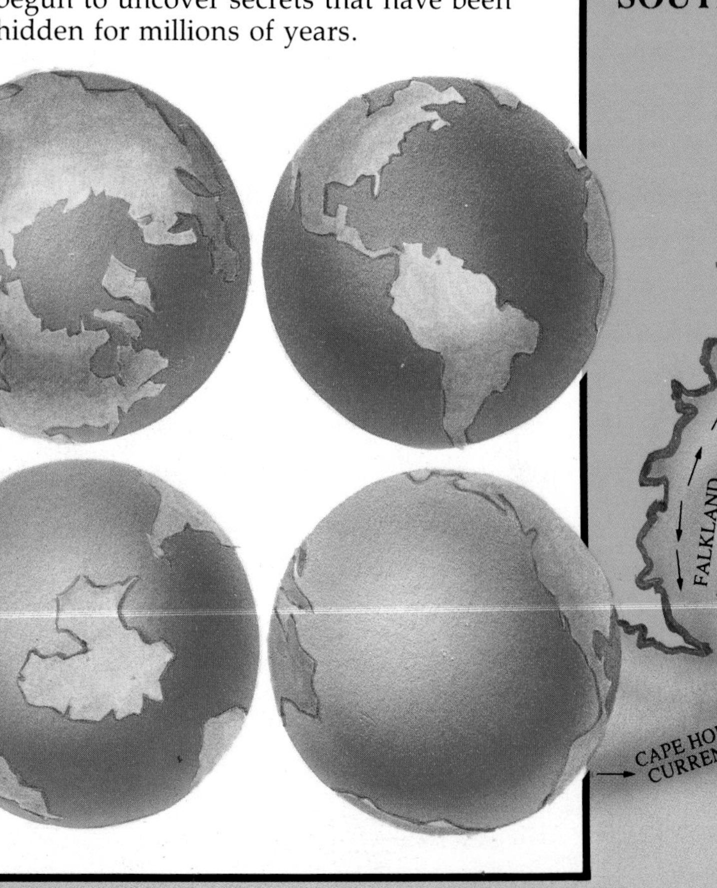

NORTH ATLANTIC DRIFT
GULF STREAM
CANARY CURRENT
ATLANTIC
NORTH EQUATORIAL CURRENT
ANTILLES CURRENT
AFRICA
CARIBBEAN CURRENT
GUINEA CURRENT
SOUTH EQUATORIAL CURRENT
SOUTH AMERICA
OCEAN
BRAZIL CURRENT
FALKLAND CURRENT
CAPE HORN CURRENT

Water in the seas is always on the move. It is influenced by wind, heat, currents, and the gravitational pull of both the moon and the sun. It is also influenced by the rotation of the earth on its axes.

Besides moving surface layers of water, wind also creates waves that range from only a few centimeters to many meters in height. One of the highest waves ever recorded was almost 34 meters high - about the height of a ten-story building.

Water moves in currents from the depths to the surface and back again. As water is heated by the sun, it rises. When it cools, it sinks. In this way, water is constantly circulating from the depths to the surface. Strong currents affect the routes that ships take. For example, the Gulf Stream, which reaches speeds of almost 4 km per hour, is used by ships to travel to Europe from the United States.

ARCTIC OCEAN

EUROPE

ASIA

Sea and ocean water is salty. It gets its salt from the land! Salt in the earth is slowly dissolved into the rivers and streams that flow into our oceans. There the salt remains because it doesn't evaporate with water. Some bodies of water are saltier than others.

Tropical waters, like the Red Sea, lose more pure water in evaporation from the sun's heat. This leaves a greater amount of salt per volume of water. The polar seas are less salty as they have less evaporation from the sun.

JAPAN CURRENT

PACIFIC OCEAN

NORTH EQUATORIAL CURRENT

Tides occur when the gravitational force of the moon and sun pulls on the earth and lifts the water. As the earth rotates, water closest to the moon is more strongly affected than water further away and a "bulge" grows in the ocean. This "bulge" expands and develops into tides. Tides occur every 12 hours and 25 minutes.
All of these forces, combined with the influence of the weather, the shape of the ocean bottom and the continents, work to create major water flows throughout the world.

MOZAMBIQUE CURRENT

WEST AUSTRALIAN CURRENT

AUSTRALIA

EAST AUSTRALIAN CURRENT

INDIAN OCEAN

WEST WIND DRIFT

WATER ON THE MOVE

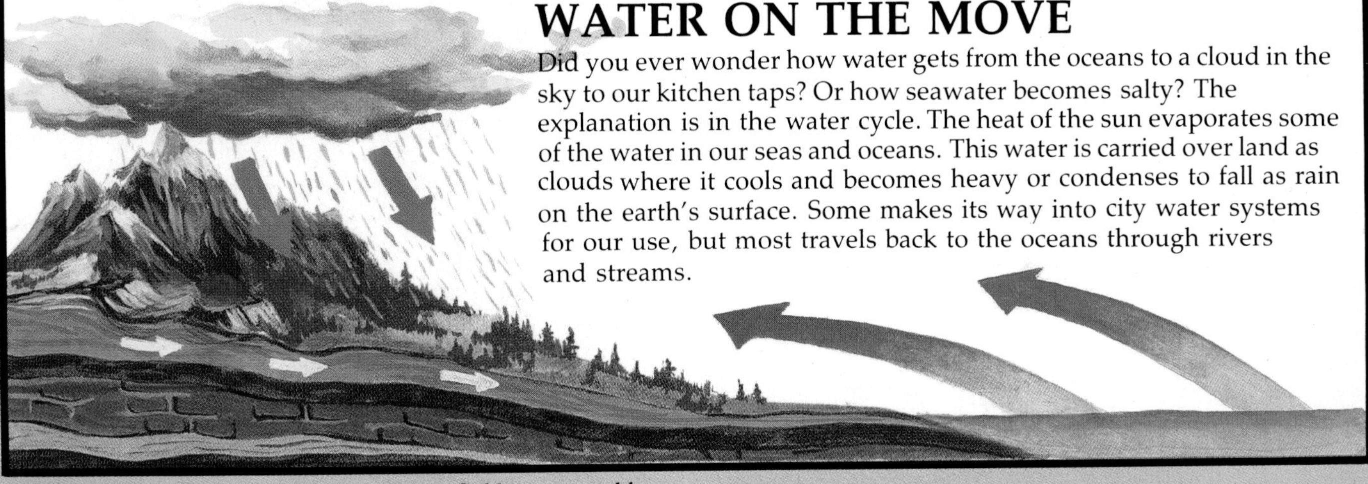

Did you ever wonder how water gets from the oceans to a cloud in the sky to our kitchen taps? Or how seawater becomes salty? The explanation is in the water cycle. The heat of the sun evaporates some of the water in our seas and oceans. This water is carried over land as clouds where it cools and becomes heavy or condenses to fall as rain on the earth's surface. Some makes its way into city water systems for our use, but most travels back to the oceans through rivers and streams.

OCEAN CURRENTS : Warm currents= red. Cold currents = blue.

CANYONS AND MOUNTAINS

In many ways, the bottom of the oceans is very similar to the land on the continents. Both have prairies, cliffs, mountains and canyons. But there is a difference. While almost every square kilometer of land has been walked on by man, much of the area beneath our seas and oceans remains unexplored.

When we go to the seashore, we usually see the land sloping gently into the water. Even where there are cliffs, there is usually a bed of broken rocks or sand at the bottom, and from that point on the bottom slopes gently out to sea. It gives you the impression that it goes on forever.
But, it doesn't!

Surrounding the continents is an area of gently sloping ocean floor extending, at some points, hundreds of miles out to sea. This area is made up of soil that has been washed off the land by wind and rain. Some of it - such as at the base of cliffs - has been gradually broken away by the action of waves. This process, called attrition, gradually breaks up the rocks into smaller and smaller particles until they become sand. Known as the continental shelf, this is the section of continent submerged under water. Since it is shallow and receives a lot of nutrients from the land, it often has an abundance of marine life. West Africa is an exception; it has no continental shelf.

The edge of the continental shelf is much steeper, sloping 500 meters or so to the ocean floor. As you can imagine, it is something like the edge of a cliff. This area is known as the continental slopes.

Beyond the continental slopes are the great expanses of ocean floor, called the abyss. In some places, it goes on for hundreds of kilometers without so much as a bump or dimple, while at other places it is mountainous, has huge canyons, and can be as deep as 11,000 meters. Some islands are simply the peaks of underwater mountains, while others have been made in recent times by underwater volcanoes that push up from the sea floor.

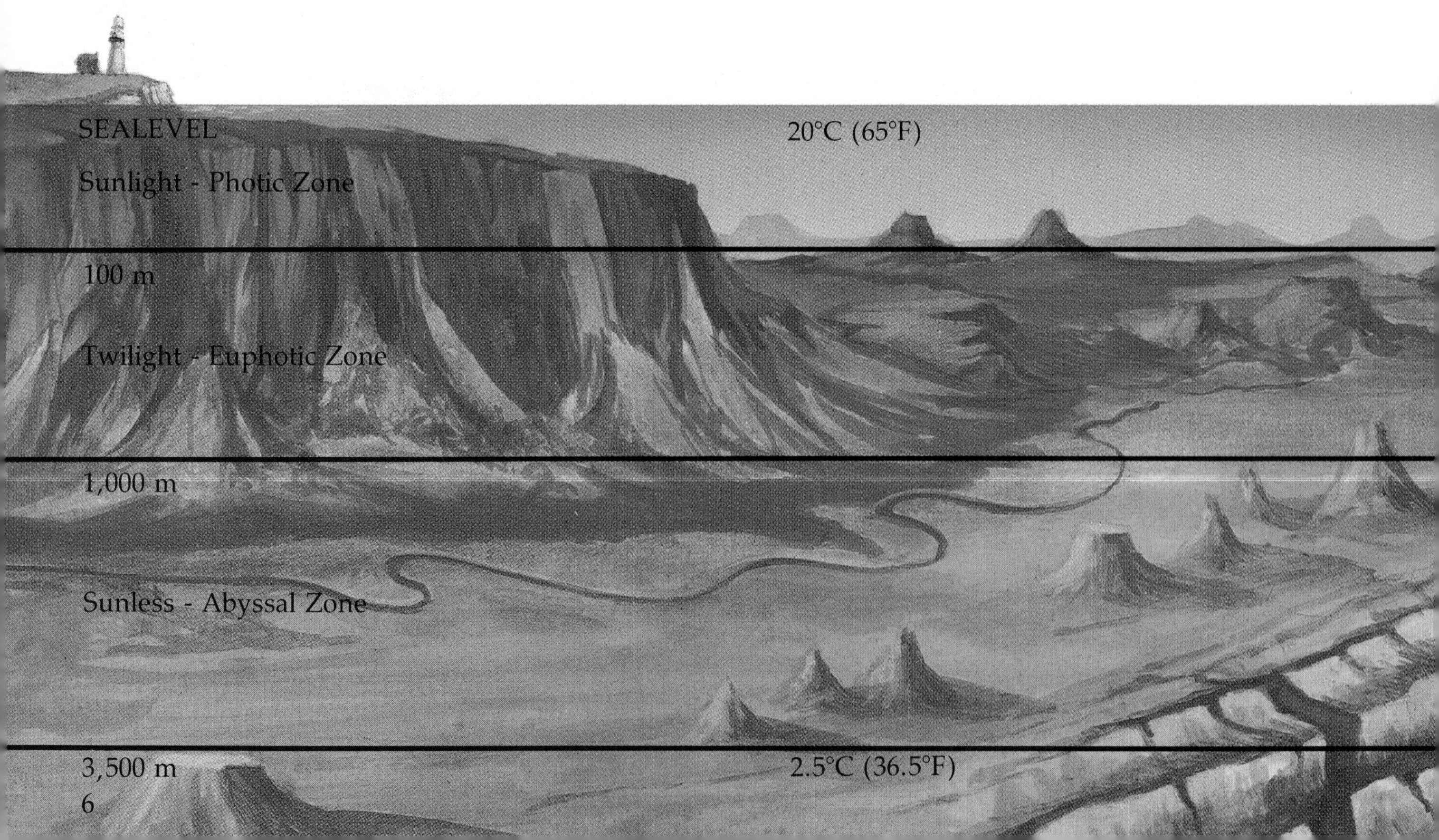

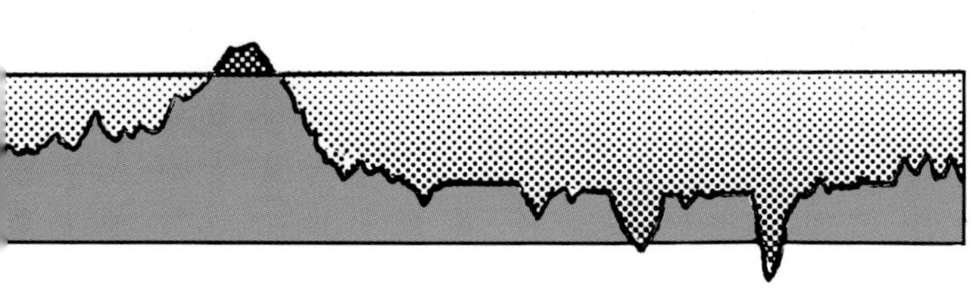

HOW DEPTH CHANGES WATER

As you go deeper into the water, it gets colder, darker, and the pressure increases.
Sunlight has a limited penetration in water because its energy is absorbed. This means the deeper you are, the less light to see with and the less heating of the water. Scientists have divided the sea into depth zones as shown on the following diagram:

Sunlight - Photic Zone	surface to 100 m
Twilight - Euphotic Zone	100 m to 1,000 m
Sunless - Abyssal Zone	below 1,000 m

In some places the surface temperature of the water is about 20°C (65° F) or warmer. Then it gradually gets colder the deeper the water until it stabilizes at about 2.5°C (36.5°F) around 3,500 meters below the surface. Water temperature near the ocean bottom remains almost uniform at 0°C (32°F) throughout the year.

PRESSURE OF WATER

As an underwater explorer, you might worry about the cold temperatures, but the pressure of the water would crush you long before you got that deep.
Let's think of pressure in terms of buckets of water, as though we have to support a meter of it for every meter we go down. You might want to pick up a full bucket of water to get an idea of how heavy it is. A bucket of water is about a third of a meter deep.

If you were to pick the bucket off the ground and then someone hung another bucket from it, and then another, you would be holding about one meter of water - or about 100 kg - as much as a heavy man! As you go deeper in the water, it's like adding buckets to your load. The human body can stand a lot of pressure, but about 100 meters is the deepest that men can dive in underwater suits.

CREATURES OF THE SEA

Life in the oceans is connected and dependent on the food chain. At the base of this chain are tiny organisms (some as small as a single cell) called plankton.

These can be plant or animal organisms. The plant organisms, called phytoplankton, feed on nutrients that are washed off the land or are brought up from the continental shelf by currents. Since plants require sunlight to exist, phytoplankton grow in a thin layer at the surface. Zooplankton (the animal organisms) feed off cells or exist as worms, buglike creatures, or miniature shrimplike animals called krill. Some zooplankton feed off others. Small fish feed off zooplankton, which in turn feed larger fish, which in turn feed still larger fish, creating what is called a food pyramid. As you might guess, there are many, many times more smaller creatures in the sea than larger, since larger creatures feed on smaller ones - usually the next smaller series in the chain. One exception to this is the whale. Some species of whale feed only on plankton, especially krill.

In this pyramid, about 450 kg of plants will support 45 kg of plant-eating animals, which will support about 4.5 kg of meat-eating animals. And this will support about 0.5 kg of human flesh.

HAMMER-HEAD SHARK

BLUE SHARK

PELAGIC FISH

Most fish with normal skeletons are called pelagic fish, and they come in millions of shapes, sizes and species. More commonly known for man's food are tuna, herring, mackerel, anchovies and salmon.

INVERTEBRATES

Creatures with no skeletons or backbones are called invertebrates. Some of them carry their own shell. Examples of this species include shellfish like clams, oysters, conches, mussels and snails.

DOLPHIN

SEA TURTLE

CARTILAGENOUS FISH

Fish with gristly, rather than bony, skeletons are called cartilagenous. These include fish like skates, rays and sharks. Cartilagenous fish do not have covers over their gills. While other fish can remain stationary and breathe by pumping water through their gills by opening and closing their mouths, the shark needs to be constantly moving for oxygen-laden water to pass through its gills. Only rarely does it remain stationary in a kind of semi-hibernation.

CRUSTACEANS

Creatures that have their skeletons on the outside are called crustaceans. Krill are crustaceans, as are lobsters, shrimp and crabs.

CEPHALOPODS

Squid, cuttlefish and the octopus are examples of cephalopods which have tentacles. They appear dangerous, but are really harmless.

SNAKES AND EELS

Sea snakes exist in the Pacific Ocean and have a deadly poisonous bite. Eels look ferocious, but most are harmless.

MORAY EEL

CORAL

Tiny invertebrates that exist in warm waters form colonies. The many varieties build their houses of calcium-like materials extracted from the sea. As they die, they leave their houses behind and these build up into coral. Although all corals are beautiful, they cause painful wounds to swimmers and divers who brush against them. Coral reefs provide protection for fish and have become such a convenient area that one-third of all fish species live in them.

MAMMALS

Warm-blooded creatures are called mammals. Those found in the sea include whales, porpoises, turtles, sea lions, seals and manatee. All breathe air and must come to the surface frequently.

CORAL REEF

HORSESHOE CRAB

LOBSTER

SKATES AND RAYS

These creatures usually lie in the sand on the sea bottom. They do not swim like fish, but "fly" in the water, propelling themselves by their wings. One exception is the manta ray which flies near the surface, feeding on krill and small fish. Sometimes the tips of their wings come out of the water and they are mistaken for sharks.

DEMERSAL FISH

Demersal fish include cod, flounders, hake and skates. Some of them can change colors to blend in with their surroundings.

FRIENDLY CREATURES AND CURIOUS MONSTERS

PORPOISES

Porpoises have long been known as man's friend. They enjoy playing alongside ships, often "surfing" in the bow wave of a ship. Many stories abound of sailors who have been saved by porpoises, and in a few instances porpoises have been known to frequent beaches, allowing children to pet and ride them. In Australia, porpoises play with windsurfers, pushing them off their boards as they swim by.

In aquariums, porpoises and killer whales compete for attention by doing tricks. Both killer whales and porpoises communicate by squeaks and clicks that can be heard by underwater microphones.

WOLF EELS

Off the coast of British Columbia, divers often find wolf eels which are the ugliest, most ferocious-looking eels in existence. Yet, they are harmless to man and will even accept a meal from the hand of a diver.

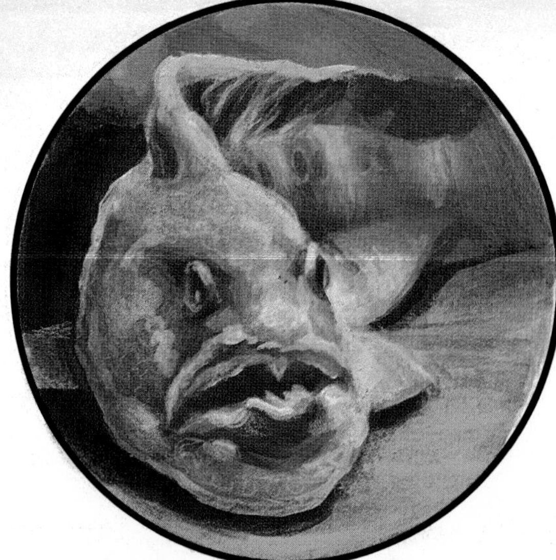

KILLER WHALES

Along the west coast of North America, where seals live on rocky shores, killer whales stalk seals as they dive in the surf for food. Sometimes these whales come right up on the rocks to drag their prey into the frothy water. Their sharp, pointed teeth can rip and tear as they feed on prey. If a man was swimming in the same waters, he might be mistaken for a seal or other large fish that killer whales like to eat.

Yet, killer whales are friendly to man! In aquariums around the world they are trained to do tricks, jump through hoops, and carry riders on their backs! They like to be petted, talked to and have been known to allow a man to put his head in their mouths! Even in the open sea, divers have been known to frolic among the killer whales, unmolested.

BARRACUDA

Barracuda can look terrifying because they have razor-sharp teeth and round, staring eyes and can be almost as long as a man. Barracuda swim extremely fast. They lie in wait, then strike a fish before it has a chance to move, usually cutting it in two with its ferocious teeth.

Knowing this can add to a diver's terror, but in truth the barracuda will rarely attack anything as large as a man. Smart divers avoid wearing shiny bracelets and rings, however, just to be sure that a sudden movement or flash of light doesn't fool the barracuda into thinking it's a fish!

OCTOPUSES

Stories of giant octopuses or squids that have captured ships or submarines and dragged them to the depths have been told for years, but the truth is, these creatures are neither huge nor dangerous. The octopus is a very shy creature and difficult to approach under water. When in difficulty, it releases a cloud of inky fluid to try to confuse its enemy while it squirts away.

MANATEES

These slow-moving creatures are sometimes called sea cows or mermaids, since ancient sailors are said to have mistaken this mammal for half-woman/half-fish. Shy, harmless creatures, they are nevertheless ugly. They feed on sea plants, munching them just like a cow.

DANGERS IN THE WATER

WIND AND WAVES

Storms at sea have claimed thousands and thousands of lives. Waves can build up to five meters high and more, easily capsizing large boats. A tidal wave, called tsunami, is caused by underwater earthquakes and can cause widespread devastation, as in Bangladesh where villages were wiped out in 1985, leaving over 200,000 homeless.
While swimming, wave action, along with offshore currents and tides, can cause an undertow that can pull you under and out to sea. Wave force can throw you against jetties, rocks or reefs, causing injury or death. When there is extensive wave action at beaches, the water is usually murky. Swimming in these waters can be dangerous. One common danger is for fish that are not normally aggressive toward humans to strike a swimmer, perhaps by mistake. Kingfish and barracuda have been known to do this.

JELLYFISH

The Portuguese man-of-war is known all over the world for its sting. These creatures float along in the sea, pushed by the wind.

Their tentacles can grow to twenty meters, and small fish that swim into these spindly arms are quickly poisoned. A swimmer accidentally touching the tentacles will suffer intense, burning pain and sometimes die from shock. Onshore winds blow these jellyfish onto the surface of the water and waves wash them up on the beach. Just stepping on one can cause severe discomfort.

GREAT WHITE SHARK

SEA SNAKE

EELS

Despite their unpleasant appearance, not all eels are dangerous. The moray eel will clamp onto an intruder's hand that has found a way into its den. In this instance, struggling is useless, and it's best simply to wait until the eel lets go. The electric eel, on the other hand, can produce a charge that can stun or even kill a swimmer.

SEA URCHINS

These plantlike creatures have spines that lodge into the skin when brushed against or stepped on and can cause pain similar to that of a man-of-war, though not nearly as severe.

PIRANHA

These small fish, found throughout South American rivers, measure an average of 30 cm long. Their razor-sharp, triangular teeth and powerful jaws can easily tear and rip flesh. They attack in schools and a creature the size of a cow can be devoured in about fifteen minutes, leaving just a skeleton and some tough hide.

RAYS

Rays lie hidden in the sand beneath the water. Except for the manta ray, all other rays whip their tails against intruders which causes ragged cuts that are slow to heal. These creatures will not attack, but whip up their tails in defense when disturbed. To avoid them or scare them away, one should shuffle, rather than step, in the water.

SEA SNAKES

One of the deadliest creatures in the world is the sea snake. They are found mainly in the south Pacific Ocean with none in the Atlantic. To be bitten is to suffer certain death.

SHARKS

Some species of sharks, like the dogfish, whale, basking and sand sharks, are harmless. However, most are dangerous maneaters. In general, it's wise to avoid sharks.

STONEFISH AND LION FISH

These fish are ugly and unappealing. When touched, they inject a poison from the spines on their backs that can cause pain and paralysis.

LION FISH

STONEFISH

CREATURES IN THE ABYSS

Many weird and unknown creatures live in the abyss. Recently, huge sea worms and crustaceans have been found around areas heated by underwater geysers. Many deep-water fish are blind, but some, like krills, produce their own light.

MAN UNDERWATER

SNORKELING

Snorkeling gives us the opportunity, through the aid of a face mask, snorkel and fins, to see the world beneath the waters.

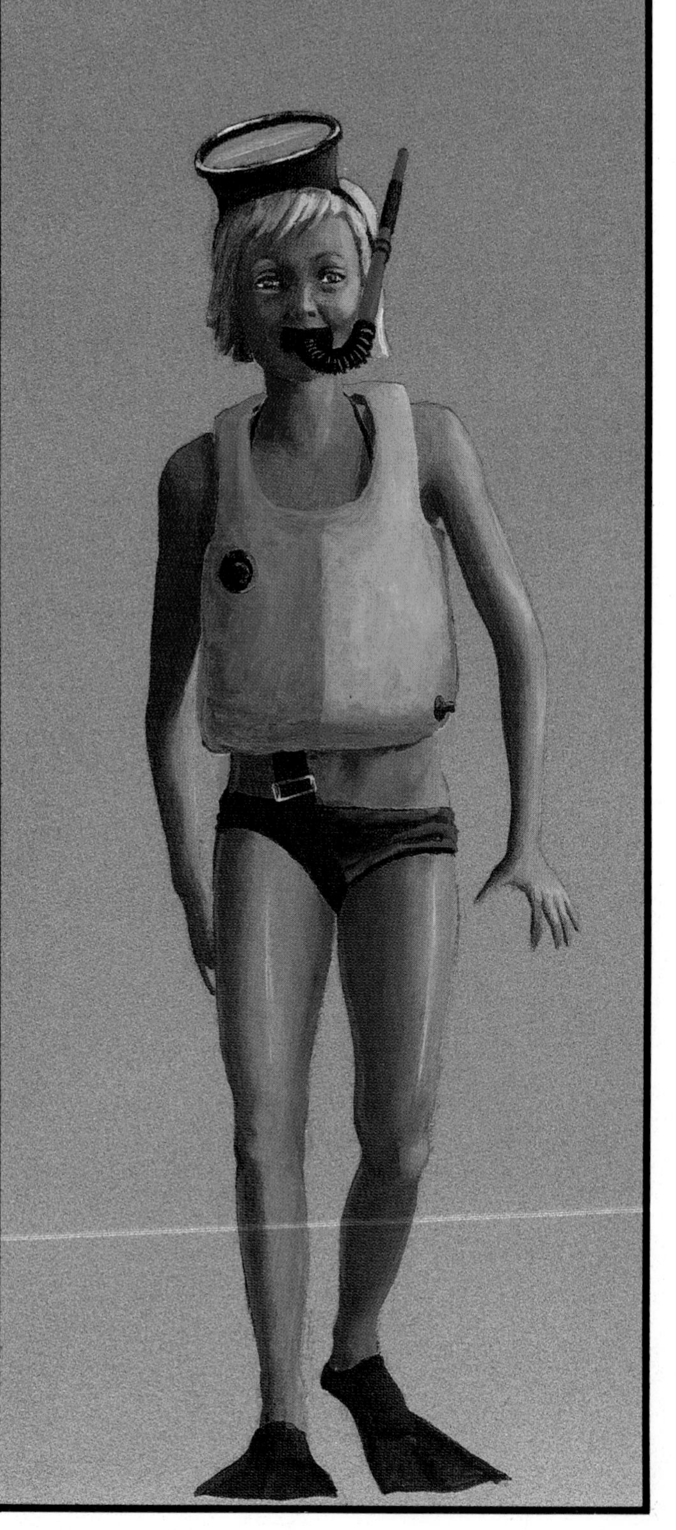

Good equipment is essential for successful snorkeling. Use a full face mask with tempered glass. Choose one that fits your face well with a good seal. You can check the seal in the shop where you buy it by holding it to your face and taking in a breath through your nose. If it will stick to your face without falling off (and without the strap on), then it will not leak in the water. Snorkels that are made of hard plastic or other materials should not be used, because if you bump into something as you are swimming, it can hurt your mouth. Get one that is made of soft rubber and attach it to your mask with a rubber strap so

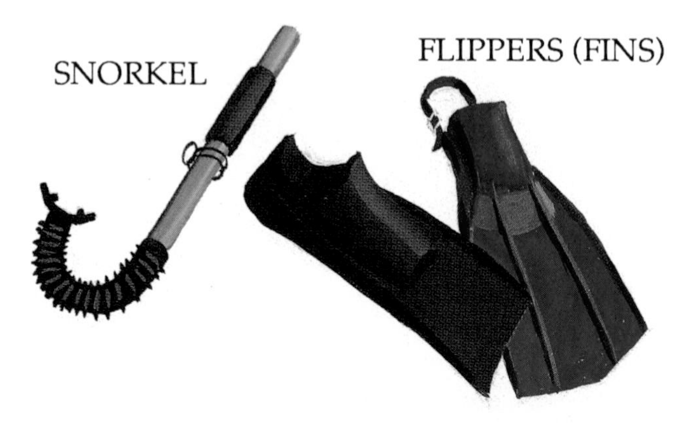

that the snorkel is on the left-hand side of your head. This is important if you intend to be a scuba diver some day!

Fins come in two basic styles: full-foot fins or open fins with adjustable straps. Open fins are preferable if you intend to scuba dive, because they can be adjusted to accommodate wet suit boots. Also, if you are still growing, you can adjust them as your feet get bigger. Be aware that some cheaper fins on the market are too flexible to give you much power. A good fin can increase the power of your leg kick four times.

Lightweight flotation vests that make snorkeling safer are available. They can save your life and should always be worn.

THE BUDDY SYSTEM

Always snorkel with a friend, never alone, so that one can help the other in an emergency. If snorkeling from a boat, it's always best to have one person stay on board to watch how everyone is doing.

DIVE FLAGS

Even when snorkeling from shore, a dive flag, flown from the boat, should be used to alert boaters to stay clear; that there may be divers under the water. When snorkeling from shore, a flag can be carried on a float or inner tube. A small anchor is used to hold it in place.

LEARNING TO SNORKEL

Learning to use snorkeling equipment should be done under supervision.
Saliva can help to prevent your mask fogging up in cold water. Simply spit into your mask before you put it on and spread the saliva around with your fingers. Commercial antifog ("artificial spit") can be bought at most dive shops if you prefer.

It takes time to overcome the feeling of claustrophobia one gets when swimming with a mask and snorkel on, but a little practice and patience in shallow water will do the trick. Once you are comfortable breathing through a snorkel, the next step is to learn to clear the snorkel after you have immersed it under water. The best method of clearing a snorkel is done with a sudden burst of air as you surface.

If water leaks into your mask, it can be forced out without coming to the surface by tilting your head back, pressing gently on the top of your mask with your finger, and blowing through your nose. You should practice this in shallow water until you can do it easily. Then you should practice in deeper water - about two or three meters deep. When you can throw your mask and snorkel to the bottom, dive down, put it on, clear the mask and then surface, you are ready to begin serious snorkeling. Do not begin snorkeling in strange waters until you can master this technique!

Jumping into the water with a mask on should be done with one or both hands holding the mask firmly against your face so that it will not come off or fill with water. When not using your mask, don't put it up on your head like you see in advertisements! A wave can wash the mask away; or other divers may think you are in danger as this is a known signal of distress. Instead, wear the mask around your neck like a collar until you are ready to use it.

DUCK DIVING

The secret to diving deep from the surface is to get a good "duck" dive. To do this, take a deep breath, tuck your head down and your knees in. Then shoot your legs straight up out of the water behind you.

SCUBA

Scuba diving provides greater freedom to travel in the undersea world. Jacques Cousteau, the French diver and explorer, and his partner, Emile Gagnan, a French-Canadian, were the founders of this freedom. In 1943, they invented the aqualung, or Self-Contained Underwater Breathing Apparatus (SCUBA).

LEARNING TO SCUBA DIVE

Many dangers associated with scuba diving can be reduced with proper training. **Never, never attempt to scuba dive without professional instruction!** In this way, diving-related accidents and deaths can be avoided. A number of national and international organizations throughout the world provide certification for all levels of diving.

When scuba diving, it is helpful to keep a log book. This should include the dates of dives, location, purpose, water temperature on surface and at various depths, water conditions (calm or rough), tank pressure before and after the dive, time under water at different depths, greatest depth achieved, and the names of your fellow divers.

AIR SUPPLY

One or two tanks strapped to your back are used to supply air for breathing. The air in the tanks is at very high pressure; as a result, the tanks must be handled carefully.

The air supply is directed through a regulator and is used in two stages. The first stage reduces the high-pressure air in the tanks to a safe breathing pressure. The second stage gives you air through the diaphragm as you require it. When you stop breathing, it stops, and when you exhale, it releases the air.

It is customary to have the mouthpiece hose arranged so that the hose comes from the diver's right-hand side, with the snorkel on the left. Some divers wear a spare regulator (called an Octopus) as a backup in case the first one fails.

UNDERWATER SUITS

Suits are used for two reasons: to provide warmth against the cold and to give protection against some of the dangers found beneath the seas. Two types of suits, the wet suit and dry suit, are available. Both are made of rubber, some others of nylon and lycra. Wet suits allow a layer of water to seep in, which acts as an insulator after it is heated by your body. A dry suit, used in combination with a type of insulated long underwear worn underneath, traps a layer of air inside and keeps you warm and dry.

WEIGHT BELTS

Lead weights, usually 5 to 10 kg, are attached to a belt around the waist since the human body tends to float, even with the diver's equipment. The belt has a quick-release buckle so that it can be removed quickly in case of emergency.

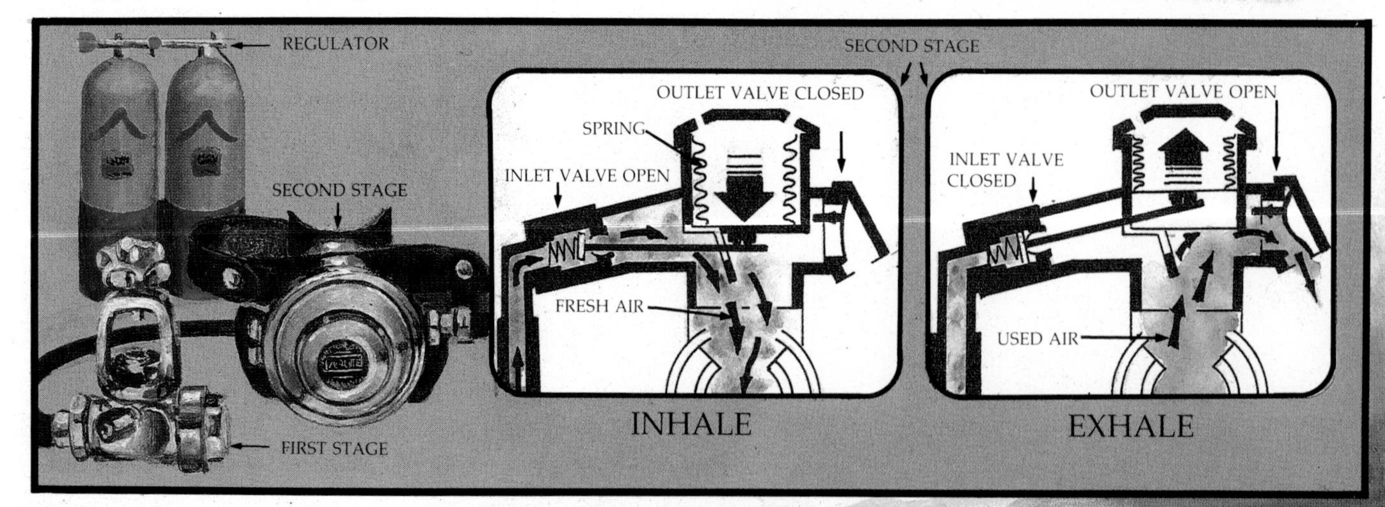

BUOYANCY COMPENSATOR

If you had to swim on the surface with all that equipment on, you would have a difficult time staying up. A buoyancy compensator that can be filled with air from your tank or by blowing into it through a mouthpiece, can keep you buoyant on the surface. A valve to release the air allows you to sink to the bottom or rise to the level you require.

Divers carry extra items as they need them, such as a knife, compass, depth gauge, pressure gauge, watch, gloves, spear guns, catch bag, flashlight and many others.

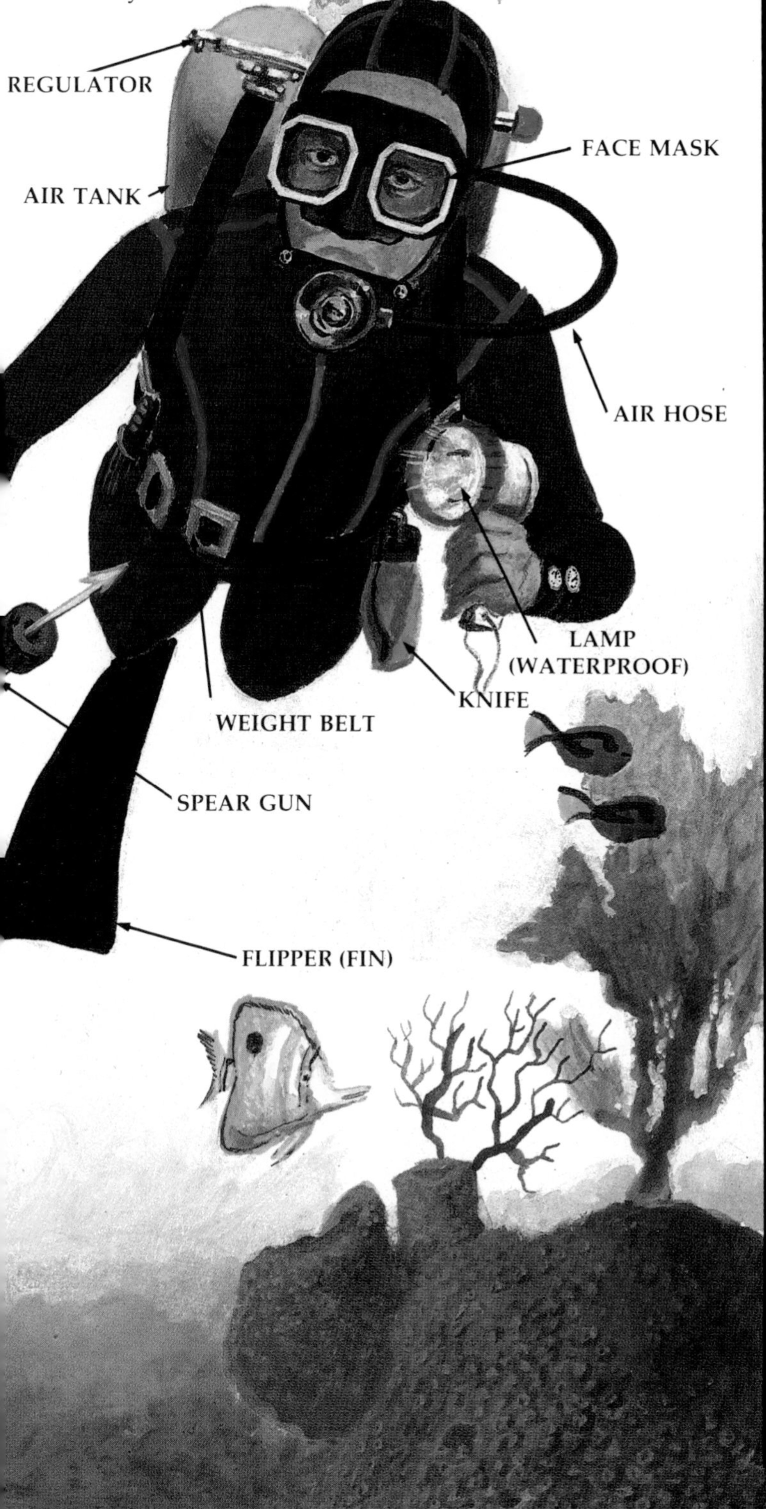

THE BENDS

When you are breathing compressed air under water, both oxygen and nitrogen are absorbed by your blood. Nitrogen is not a problem when you're on land because there isn't enough pressure to force it into your blood, and it simply goes in and out of your lungs. The problem under water is that it takes time to be released from your bloodstream. When you surface, the nitrogen forms into bubbles in your bloodstream, like the fizz when you open a soft drink. This can cause terrible pain, even death. It is called the bends because the bubbles usually collect in your joints, which causes you to double over. Divers calculate how deep they can go and for how long without getting into trouble. Emergency centers for divers with the bends have a decompression chamber, where they are put into a tank that is under pressure. Then, slowly - over a matter of hours - pressure is reduced so that the nitrogen is released a little at a time without forming into bubbles.

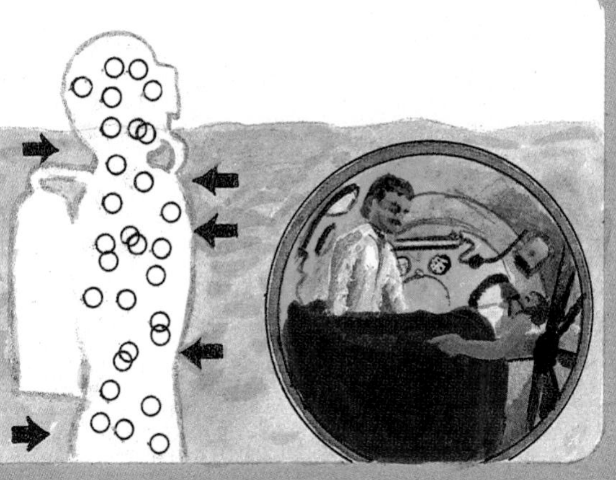

EMBOLISMS

When diving, the pressure on a diver squeezes him and everything around, including the air he breathes. If he held his breath and moved too quickly to the surface, the air would expand like a balloon. It could move from his lungs through his bloodstream in the form of bubbles. These bubbles could travel to his heart or brain and cause instant death! This condition is called an embolism. Divers are trained to rise to the surface very slowly, letting the expanding air out of their lungs and never, never holding their breath.

FOOD FROM THE SEA

WHERE ARE FISH CAUGHT?

Most fish populations live on the continental shelves, particularly in areas where there are offshore breezes and upwelling (underwater currents) that bring nutrients from the ocean floor. One example is the Grand Banks off the coast of Newfoundland. Another is the Great Barrier Reef off the east coast of Australia.

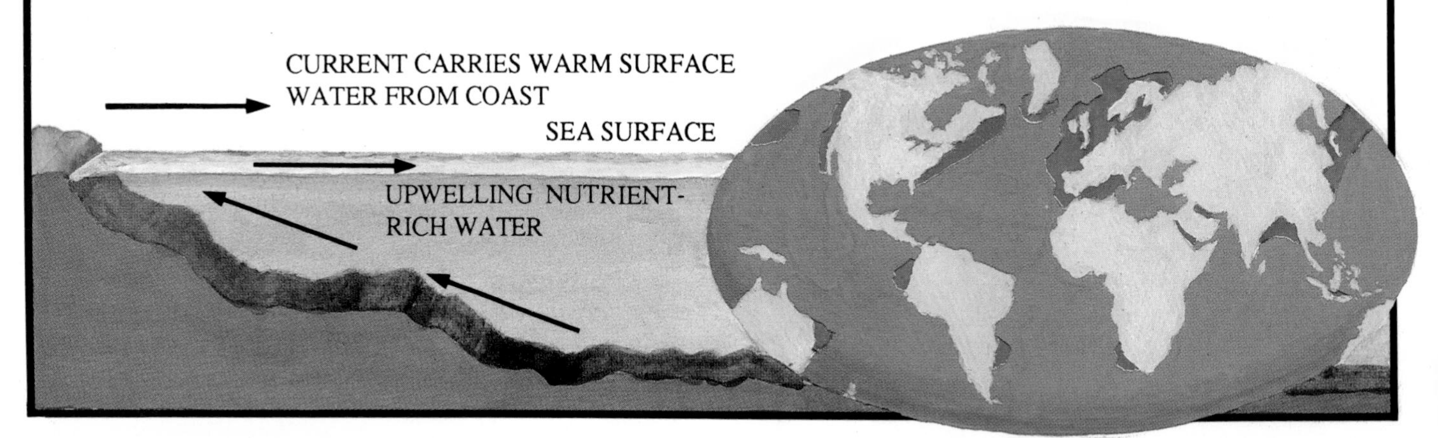

CONTINENTAL SHELVES MARKED IN RED

HOW MUCH FISH ARE CAUGHT?

The amount of fish caught by mankind has increased dramatically. In 1950, about 18 million metric tons of fish were caught around the world. Today, that number exceeds 63 million metric tons.

WHO DOES THE FISHING?

In the past, people fished off their own shores, but now fishing boats from major countries can be found almost everywhere. Fishing boats can travel far and wide because of modern refrigeration, since they can freeze the fish on board. Some countries, such as the Soviet Union, have factory ships that process and freeze fish caught by smaller boats. Japan and the Soviet Union catch more fish than other countries.

WHAT KIND OF FISH ARE CAUGHT?

Of the five basic categories of fish caught, over 50% are pelagic fish, such as salmon, tuna and mackerel. About 35% are demersal fish, such as cod and haddock, 10% are shellfish, and the rest are cephalopods (octopus and squid) and mammals such as whales and seals.

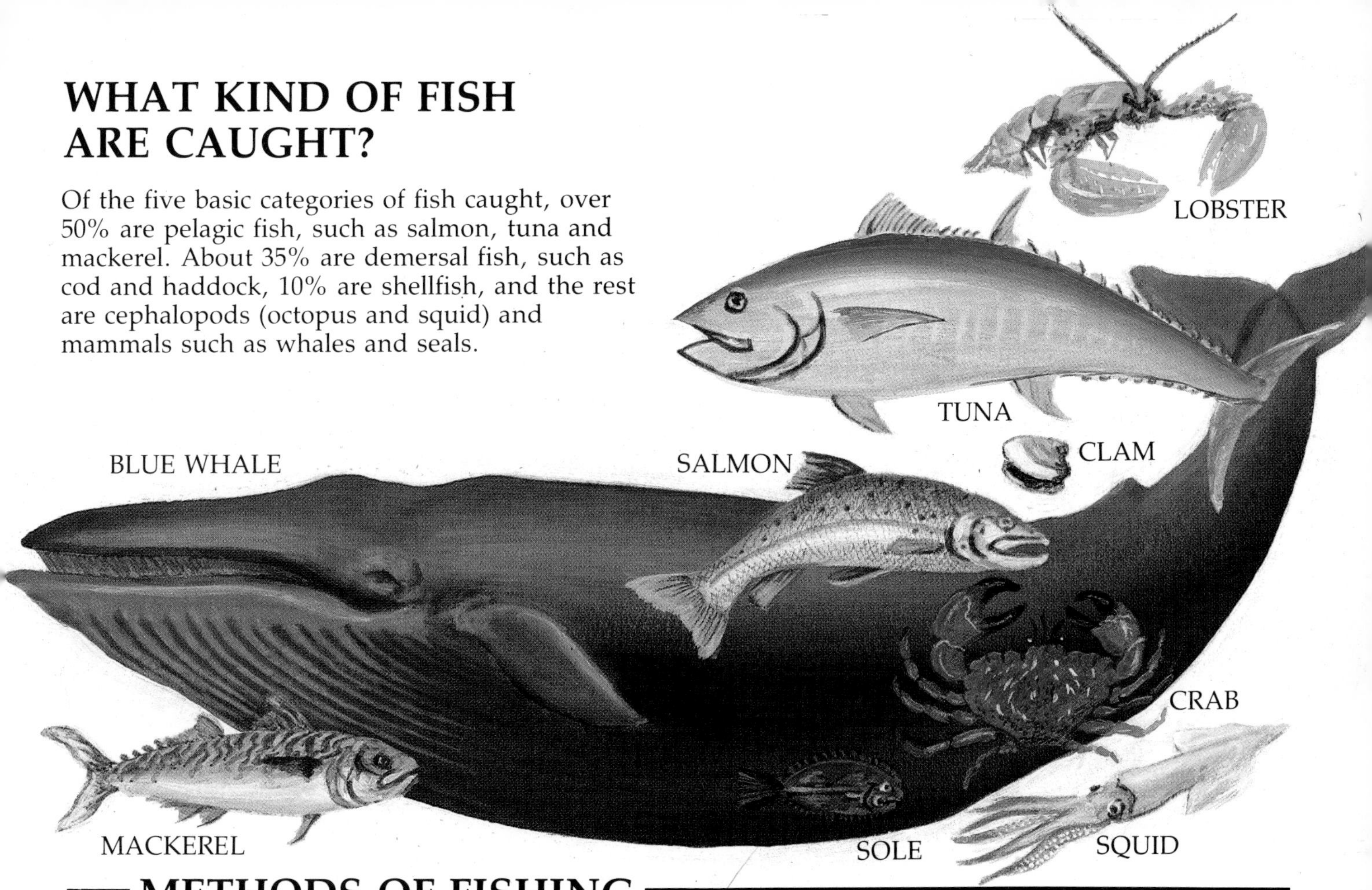

LOBSTER
TUNA
CLAM
BLUE WHALE
SALMON
CRAB
MACKEREL
SOLE
SQUID

METHODS OF FISHING

DRIFT NETS

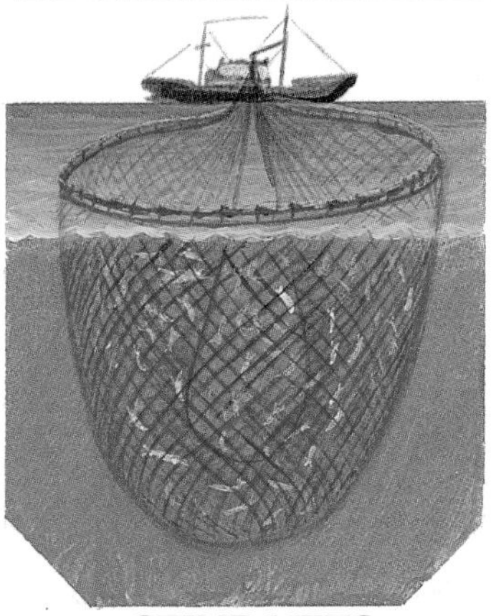

SEINE NETS

TRAWLER NETS

TRAPS

Lobsters and crabs are caught in traps that are baited with a piece of fish. These crustaceans can get into the trap to feed, but can't get out.

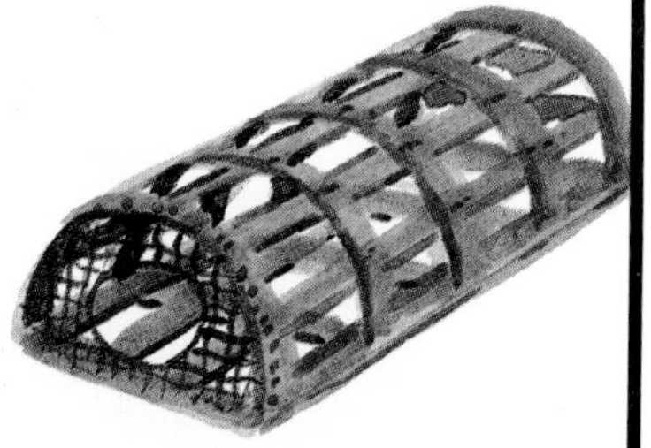

MANAGING THE OCEANS

Many problems have resulted from overfishing. The whale population, for example, has been reduced by over 90%. Most countries have banned whaling, but some have refused to cooperate. Fishing by larger countries off the shores of smaller countries has also depleted their stocks, making it difficult for people who rely on local fish to survive. West Africa, for example, is suffering from a reduction in available fish. Pollution of our oceans is becoming a problem as ships continue to dump oil and chemicals into the sea. Even more of a problem are human and toxic wastes from the land that destroy plant and animal life in the estuaries along the coasts.

FARMING THE SEA

Farming of water plants and animals for use by man is known as aquaculture. The idea of artificially farming the oceans and seas is becoming more important as the natural growth of sea life is used up and man overruns his fertile farming land with cities and highways.

OYSTER BEDS

Probably the oldest form of aquaculture involved the pearl trade. Oysters make pearls when a bit of sand or grit enters their shells. They surround it with pearl in order to prevent irritation to their soft bodies. Divers began placing sand bits into oysters, and the pearls that developed were called cultured pearls. More recently, oyster farms have been developed to provide food. Oyster farms use underwater racks as well as traditional Japanese pearl and lantern nets. Both methods allow a greater number of oysters to be cultivated over a small area.

PEARL

FRESH WATER PONDS

Carp and trout are raised in ponds where the fish are fed to increase their growth. In cold northern waters, rainbow trout grow to about 280 grams in a season. Research has shown that it's possible to increase this to 4.5 kg per season with certain species if the water temperature is controlled at 15°C all year round. Perhaps, waste heat from industrial processes could some day be used to keep pond water warm in winter, producing huge fish.

SALMON

Salmon farming begins with salmon being hatched in ponds. After being raised to fingerlings (about as long as a man's finger), the fish are released into streams and lakes. Huge salmon have been grown this way, and although the purpose has been primarily to provide sport fishing, the idea has been expanded to produce a commercial harvest.

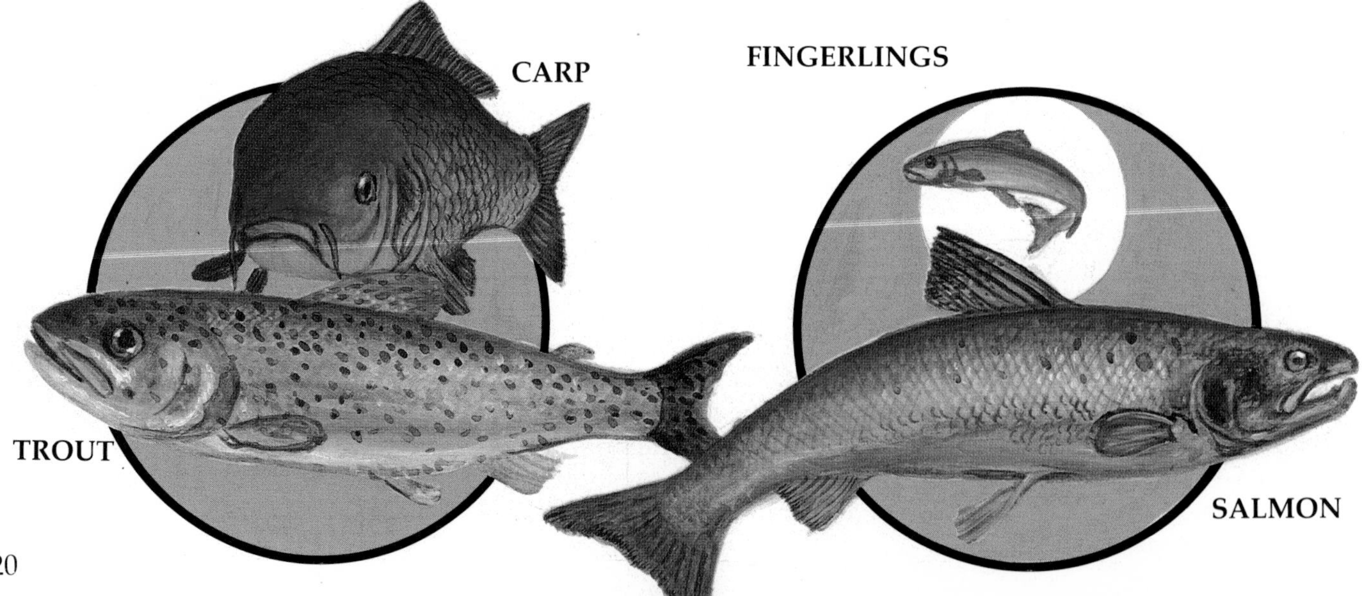

FARM-RAISED SHRIMP

SHRIMP

RED CORAL

Shrimp farms have sprung up in many areas. This member of the crustacean species has become such a popular food, natural supplies can't keep up with the demand. Egg-carrying females are cultivated in saltwater tanks where they produce 300,000 to 1,000,000 eggs. The surviving eggs develop in stages from larvae to adult shrimp. These farmed shrimp are considered tastier than those grown naturally at sea.

CORAL

Coral has long been regarded as the "flower of the sea." The 2,500 species of coral provide a wide collection of color and form. Craftsmen and artists began, centuries ago, to transform coral into jewelry and art objects to develop a multi-million-dollar business. The popular black, red, pink and gold corals are being used up. As a result, conservation of coral has become necessary.

OTHER FARMING

Japan is very active in farming carp, yellowtail and other species. This is done in narrows and shallow areas around the many islands of Japan. Kelp or seaweed is also farmed for use as food, fertilizer, and in chemicals.

A SOURCE OF PROTEIN

Fish are considered superior to meat as a source of protein. Today, Japan and Russia rely on fish for as much as 60% of their protein diet. Other populations' needs for fish are growing at an alarming rate. While this poses a threat to the wild fish population, it also presents an opportunity to create a strong industry out of one that is in its infancy.

SUBMARINES AND ROBOTS

MILITARY SUBMARINES

Submarines were invented and developed for warfare. They are becoming larger, faster and more deadly every day. To give you an idea of the size of the largest submarine, imagine standing at home plate in Yankee Stadium. The longest submarine would stretch farther than the wall at center field! If you are a soccer or football fan, imagine a submarine that is almost two fields long.
Submarines are powered by one or more diesel engines or by nuclear reactors. Electricity is generated by the diesel engines or by a steam-driven turbine with the steam produced by nuclear power. In either case, the propeller is normally driven by an electric motor. A diesel engine needs to breathe air, so when it's submerged the propeller is driven by batteries. A diesel-powered submarine can cruise just under the surface, extending a snorkel out of the water so the engines can breathe. The nuclear engine does not require air to run, so it can stay under water for very long periods, returning to port only for food and a change of crew. Air for the crew is recycled and regenerated.
Like ships, submarines remain floating because they weigh less than the water they displace. When the submarine is on top the water, it weighs less because tanks inside it are filled with air. To submerge, the tanks are filled with seawater, which increases its weight and causes it to sink. If the submarine wants to surface again, it forces the water out with compressed air.
When a submarine is under water, it can maneuver up and down by using fins on the front of the ship that, when moved, deflect the nose up or down.
Submarines have always used torpedoes for armament, but many modern submarines have missiles that can be launched both from the surface and from under water.
Detecting and locating submarines is a constant job for maritime patrols. Most submarines can be found by listening with sophisticated sonar devices lowered into the water from ships or helicopters.

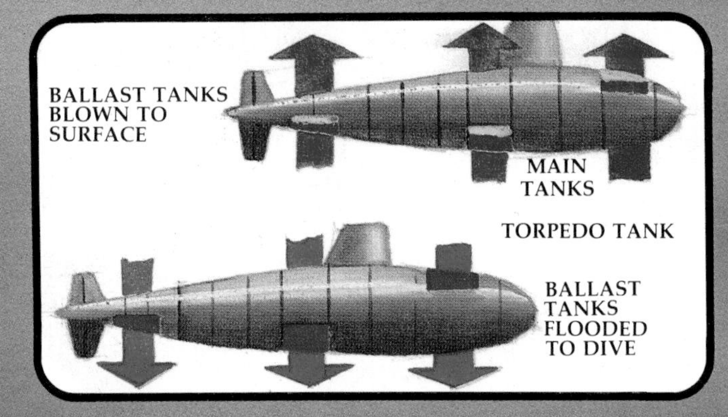

PEACEFUL SUBMARINES

Submarines are playing an increasing role in the scientific investigation of the undersea world. It's nice to see that something invented for warfare is being applied to peaceful uses.

DID YOU KNOW?
The USS *George Washington* (1959) was the first submarine to fire a missile from beneath the surface of the sea.

ALVIN AND DEEPSTAR

These two deep-water submarines have traveled to depths of 4,000 meters where the pressure of the water is about 0.4 metric tons per square centimeter. This weight would be like covering the submarine with quarters and having a big Mack truck sit on each quarter. Alvin carries a pilot and scientific crew of two. Deepstar was invented under the direction of Jacques Cousteau.

THE WASP

Named for its resemblance to a wasp, this yellow Canadian-built suit is really like a submarine that a man wears. A man inside can control its arms with great precision. Dr. Joseph B. MacInnis, a Canadian doctor who specializes in the medical aspects of diving, used the suit to explore the *Breadalbane*, a British ship that sank in the Arctic in 1843.

TRIESTE II

In 1960, Jacques Piccard and Lt. Don Walsh of the U.S. Navy piloted a bathyscaphe, Trieste II, 10,900 meters into the ocean. That's deeper than Mt. Everest is high! Not only that, they saw fish living at those depths!

1. ENGINE ROOM
2. REACTOR ROOM
3. MISSILE ROOM
4. MISSILE CONTROL CENTER
5. NAVIGATION ROOM
6. GYRO ROOM
7. STORES
8. BATTERIES
9. BRIDGE
10. PERISCOPE ROOM
11. CONTROL ROOM
12. CREW'S QUARTERS
13. CREW'S MESS
14. OFFICERS' WARDROOM
15. FORWARD TORPEDO ROOM

MODERN USES OF UNDER-WATER TECHNOLOGY

Advances in underwater technology over the past ten years have been enormous. Submersibles and robotic submarines have made the world beneath the seas easier to reach than ever before.

FLIGHT 182

Underwater technology was urgently needed to recover voice recorders and the "black box" from the Air India 747 jetliner Kanishka after it crashed on June 23, 1985, killing all 329 passengers on board.

The plane was traveling to Heathrow Airport, London, England, from Montreal, Canada. The final report by radio to Shannon Airport gave the position of the plane at 31,000 feet. Shortly after this radio message, the jumbo jet disappeared from radar screens.

Three principal theories were put forth as reasons for the crash: A bomb smuggled on board had exploded, massive structural failure had occurred, or total electrical failure took place. To find out the cause, the wreckage would have to be found in 2,000 meters of water!

The Scarab I, a robotic submarine with video cameras, was hung by cable from a French ship. It located pieces of the wreckage scattered over a four-mile area. It found and retrieved the voice recorders and "black box" of the airliner (both about the size of a chair), buried in huge expanses of ocean floor.

Next, Scarab II was used to attach cables to the pieces so they could be brought up and placed aboard ships for examination. Although there were some signs that might indicate an explosion, no conclusions were drawn about the cause of the crash.

EXPLORING THE TITANIC

The robotic submarine again displayed its importance when used by leader Robert Ballard of a French-American exploration team to locate the *Titanic*.

This luxury liner, thought unsinkable because of its mammoth size and design, struck an iceberg on its maiden voyage April 14, 1912, and sunk. With some of the wealthiest people in the world on board, it was thought that the ship's four safes would contain millions of dollars in jewelry and other valuables. The problem was how to find the ship lying somewhere on the bottom of the ocean.

This problem was solved when Robert Ballard found and photographed the *Titanic* in 1985. It was lying on the the sea floor, southeast of Newfoundland, Canada, in remarkable condition.

Using Alvin, a free floating submersible with room for three men, and an unmanned sub, Jason Jr., the team explored the massive wreck. They found china and bottles of wine undamaged as well as broken bed frames, a bathtub, and a bronze statue of the Roman goddess, Diana. In hard-to-reach places, the team used Jason Jr., controlling it with a console held by the pilot inside Alvin.

During twelve dives, the explorers were able to take hundreds of photographs because of high tech assistants -- Jason Jr. and Alvin. This combination of man and machine made the exploration of the *Titantic* possible and has given us pictures and knowledge of how seawater preserves.

UNDERWATER TREASURES

Thousands of shipwrecks all over the world have never been found. Some have massive treasures of silver and gold, such as the wreck of the Spanish galleon *Nuestra Senora de Atocha* which sunk in 1622 in a hurricane off Key West, Florida. Old sailing ships often carried gold and silver for trading, along with cargoes of goods that may have been commonplace in their day, but are considered very valuable by collectors today. The *Atocha*, which was found in 1985 by Mel Fischer after a sixteen-year search, has a cargo that is estimated to be worth $400,000,000!

To find one such wreck may be to become rich, but not all riches are silver and gold. Archaeologists and historians find artifacts just as exciting and rewarding - cannons, anchors, statues, bottles, urns - anything that gives a picture of how mankind lived in that era.

People who look for wrecks generally begin in libraries, public archives, museums and private records, not the open sea. Ancient charts, shipping records, manifests, books, letters - these provide the clues that lead searchers to the likely routes taken.

Then, hours, days, weeks, years at sea are spent scanning likely areas, reefs and ancient storm centers, anything that could narrow down the hunt. Bottom areas are examined with sonar, which is like underwater radar, to provide bottom profiles and identify strange and unusual soundings. Having found something strange or an

Marine researcher charts finds on location.

Diver collecting samples.

expected site, the floor is scanned by metal detectors, operated by a remote-piloted vehicle or sometimes hand-held by divers.
You would expect that once finding a relic or other clue, a frenzy of digging would develop. But care must be taken to identify the limits of the area to be searched, because ships that have been underwater for years break up easily and can be scattered by wind and wave action during storms. One must be careful not to disturb or accidentally bury other relics!
A wreck site is photographed and staked out in a grid with each square marked on a map. Sometimes scaffolding or ropes are used to mark the grid on the sea bottom.
Once the limits of the wreck are identified and staked out, each square is picked over with care. Suction devices (like underwater vacuum cleaners) remove the silt and sand.
Divers can stay down for about an hour without surfacing. If they want to communicate, they have to return to the surface. To avoid this, a dome filled with air is sometimes supplied, where divers can rest and communicate with others.

AIR BAGS OR WINCHES

Bags of air are used to float heavy objects up from the bottom, or a winch from the support boat can also be used to lift objects.

Divers are limited to the length of time they can remain underwater without risk of suffering the bends. An alternative is to have a decompression chamber available. This is a tank that is filled with compressed air. The diver goes into the tank where he remains while the pressure is gradually lowered over a period of time until it equals atmospheric pressure. In this way, nitrogen is slowly released from the blood instead of forming bubbles - and the bends!

Air Lift - basically an oversize vacuum cleaner.

Researcher exploring ravaged wreck of clay jugs once filled with supplies.

Metal grid permits precise location of each find of a wreck.

MINERALS, OIL AND POWER

Mankind is looking to the sea for more than just fish. There are riches in the ocean waters worth harvesting. When you consider that only 30% of the earth is land from which most of our resources are found, doesn't it sound intriguing to imagine how much more there must be under the sea, just waiting to be discovered?

MINERALS

It may surprise you to learn that man has gotten minerals from the ocean waters for more than eighty years! Salt has long been harvested from the sea. Ocean water is pumped into drying beds where it is allowed to evaporate, leaving its salts behind. These are raked and cleaned on screens to take out the debris, then bagged for use. If all the salt in the oceans was gathered up, there would be enough to cover the entire globe several times over.

Magnesium metal is obtained from the sea. A rock called dolomite is heated up and then reacted with seawater. After further treatment, the end result is magnesium oxide, which can be used to make firebricks, pharmaceuticals and magnesium metal. Recently, miners have been excited by finding manganese nodules on the floor of the oceans. Schemes have been developed for mining and separating these nodules from ships that are anchored above. Mining at depths as much as 5,000 meters has been proposed.

Magnesium is collected in shallow settling ponds.

One pound of magnesium can be extracted from 142 gallons of water.

Naturally, there are other metals on the ocean floor, such as gold, silver, lead, iron, zinc, copper and others, but mankind's needs for these have not been strong enough to overcome the high cost of underwater mining. It is nice to know that there is so much to be discovered in case our land-based resources begin to run out.

OIL

As the world uses up its petroleum resources, oil companies are looking to the sea more and more for exploration. Most of the technology for oil extraction from the sea has been developed in Texas, where oil wells have been operating in the Gulf of Mexico for many years, and the North Sea off Britain and Norway.

Some drilling for oil is done from ships that have a drill tower amidships. Other types of oil rigs include floating rigs, rigs that have massive concrete bottoms that sit on the continental shelf, and others that have a tower that is guyed by underwater cables.

Schemes have been proposed for underwater production at greater depths. These units would likely sit on the bottom and operate on nuclear power.

POWER

As a source of power, the ocean has been tempting man for many years, but only recently has mankind begun to tap this resource.

Tidal power has long been contemplated at the Bay of Fundy on the east coast of Canada, where tides are as high as ten meters. Seawater rushing through a penstock could power turbines as the tide comes in and later as it goes out, which would turn generators to make electricity. The problem is what to do when the tide is neither coming in nor going out. Engineers propose a series of dams and storage ponds to even out the flow.

Two methods of extracting power from waves have been proposed. These are the Lanchester Clam from Great Britain and the Dan Atoll from the U.S.A.

Another proposed method uses the fact that the water in the ocean is colder at depth than it is on the surface. By using water at the surface to boil ammonia (since it boils at a very low temperature), then condensing it by pumping cold seawater up from deep beneath the surface, the ammonia can be used to power a turbine. This apparatus is known as the Ocean Thermal Energy Conversion unit (OTEC). It works somewhat like a refrigerator run backward (some refrigerators use ammonia).

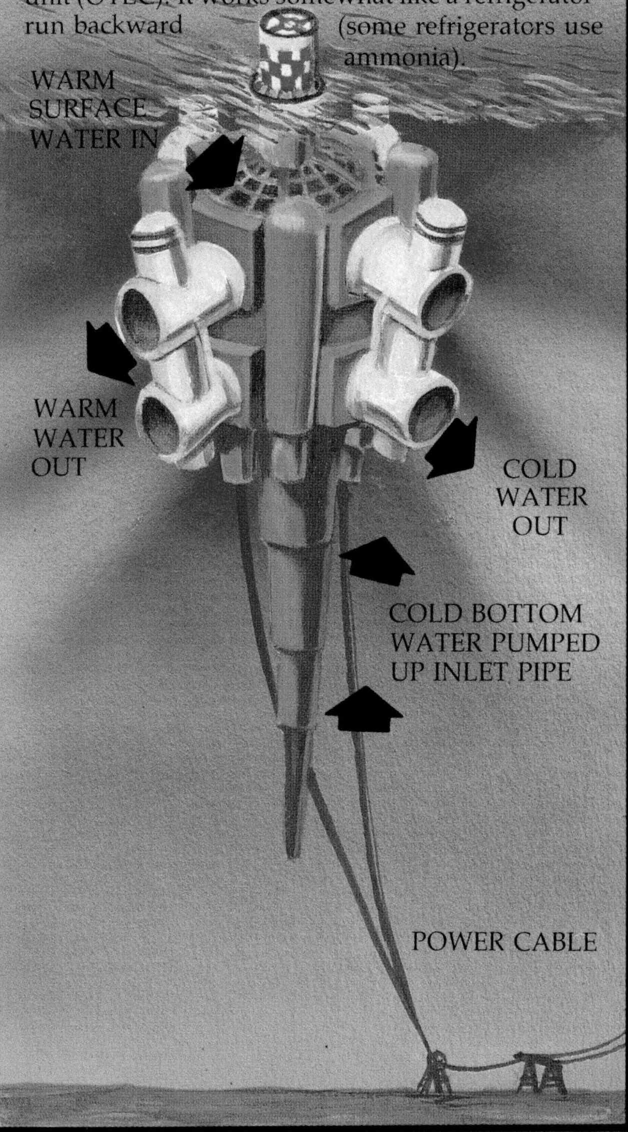

FUTURE UNDERWATER WORLDS

The underwater world holds as much excitement and as many unknowns as outer space. Exploring and using it to benefit man is as much a challenge to your generation as outer space. Mankind needs the world under our waters for food, minerals and oxygen. Future explorations of our seas will give us greater resources and opportunities for development.

MINING

Mine shafts on land go many miles underground. Is there any reason why we can't do the same from the sea? Even today, mining extends from land to ore bodies that lie under our oceans. Perhaps flexible "shafts" could extend down from anchored ships or platforms to a concrete, sealed collar on the ocean floor. In the meantime, plans exist to vacuum manganese nodules from the seabed, pumping them into the hold of a ship.

UNDERWATER CITIES

If plans to mine and explore the undersea world require that people spend a lot of time beneath the surface, it makes sense to build cities under water. Jacques Cousteau began researching this idea in 1962 and built the "Conshelf," a house where men spent over a week under the sea off Marseilles. Future worlds will have schools, hospitals, recreation - most of the same things that we have on land.

UNDERWATER RECREATION

Most people who want to see the underwater world do so by scuba diving, riding in glass-bottomed boats or by simply visiting an aquarium. Touring submarines are also becoming popular. Underwater hotels and restaurants, built alongside artificial or natural reefs, could provide tourists with a view out their windows that would be spectacular and ever-changing.

People staying in these hotels who want to have a closer look could don scuba gear and swim into the deep through an outlet in the hotel swimming pool! The scuba gear will be much lighter and less cumbersome than what is used today, however, and will include a set of artificial gills so that the oxygen you breathe is taken directly from the seawater - just like a fish!

ARTIFICIAL REEFS

Wrecked ships off the coast of Florida have become artificial reefs, attracting a myriad of fish. If this idea is exploited, then intentional artificial reefs could increase the fish population by giving coral a chance to grow and fish a place to live. After all, one-third of all our fish are found in reefs today!

SHIPPING

Nuclear-powered submarine freighters will ply the waters under the Arctic Ocean. While these will initially be used to open up the Northwest Passage, which in turn will make the Panama Canal less useful, shippers will find that the submarines can't be hurt by storms at sea. They will also find that more use can be made of a ship's volume if the ballast is cargo, especially for such things as grain, mineral concentrates and oil. As a result, underwater ports will be built so that unloading can begin under the sea. As cargo is put in, it begins to sink and is ready to sail.

UNDERWATER TRIVIA

1. Who helped Jacques Cousteau develop the aqualung?
2. What is an embolism?
3. Which diving apparatus looks like an insect?
4. Are all sharks dangerous?
5. Killer whales eat fish and seals. What do other whales eat?
6. What is kelp?
7. What is a fingerling?
8. How deep is the ocean?
9. What is a tsunami?
10. Under what conditions will barracuda attack man?
11. Is a Portuguese man-of-war a boat?
12. Where did the *Titanic* sink?
13. Are there any sea snakes off the coast of Florida?
14. Can a stingray sting?
15. What is upwelling?
16. Are plankton plants or animals?
17. What is the abyss?
18. Can you name three types of fish nets?
19. Are there mermaids in the sea?
20. What is more dangerous: an eel or a sea snake?

FURTHER READING ON THE UNDERSEA WORLD

Reader's Digest Book of the **Great Barrier Reef.**

David Miller. **Submarines.**

Louis Wolfe. **Aquaculture - Farming in the Water.**

Burgess. **Ships Beneath the Sea.**

University of Oklahoma. **Energy Under Oceans.**

Canada Center for Inland Waters. Burlington, Canada.

Communications Center. Ottawa, Canada.

Jacques-Yves Cousteau. **The Ocean World of Jacques Cousteau.**